HAMISH HAMILTON LTD

Published by the Penguin Group
27 Wrights Lane, London W8 5TZ, England
Penguin Books USA Inc, 375 Hudson Street, New York, New York 10014, USA
Penguin Books Australia Ltd, Ringwood, Victoria, Australia
Penguin Books Canada Ltd, 10 Alcorn Avenue, Toronto, Ontario, Canada, M4V 3B2
Penguin Books (NZ) Ltd, 182-190 Wairau Road, Auckland 10, New Zealand

Penguin Books Ltd, Registered Offices: Harmondsworth, Middlesex, England

First published in Great Britain 1994 by Hamish Hamilton Ltd

Text copyright © 1994 by John Yeoman
Illustrations copyright © 1994 by Quentin Blake

1 3 5 7 9 10 8 6 4 2

The moral rights of the author and artist have been asserted

British Library Cataloguing in Publication Data
CIP data for this book is available from the British Library

ISBN 0-241- 00245-1

Printed in Italy by L.E.G.O.

John Yeoman and Quentin Blake

The DO-IT-YOURSELF House that Jack Built

Hamish Hamilton

London

This is the house that Jack built.

This is the malt
   that lay in the house
      that Jack built.

This is the rat
   that ate the malt
      that lay in the house
         that Jack built.

This is the cat
  that killed the rat
    that ate the malt
      that lay in the house
        that Jack built.

This is the dog
   that worried the cat
      that killed the rat
         that ate the malt
            that lay in the house
               that Jack built.

This is the cow with the crumpled horn
that tossed the dog
that worried the cat
that killed the rat
that ate the malt
that lay in the house
that Jack built.

This is the maiden all forlorn
 that milked the cow with the crumpled horn
  that tossed the dog
   that worried the cat
    that killed the rat
     that ate the malt
      that lay in the house
       that Jack built.

This is the man all tattered and torn
that kissed the maiden all forlorn
that milked the cow with the crumpled horn
that tossed the dog
that worried the cat
that killed the rat
that ate the malt
that lay in the house
that Jack built.

Well, we only have to find out where this goes and we're nearly finished.
He's luckier than I was.
All right. If I can have a go with the power-drill.
My angel! Will you marry me?
Anyway, I might as well paint it red.
Nicely crumpled.
More trouble.

This is the priest all shaven and shorn
that married the man all tattered and torn
that kissed the maiden all forlorn
that milked the cow with the crumpled horn
that tossed the dog
that worried the cat
that killed the rat
that ate the malt
that lay in the house
that Jack built.

This is the cock that crowed in the morn
that waked the priest all shaven and shorn
that married the man all tattered and torn
that kissed the maiden all forlorn
that milked the cow with the crumpled horn
that tossed the dog
that worried the cat
that killed the rat
that ate the malt
that lay in the house
that Jack built.

This is the farmer sowing his corn
  that kept the cock that crowed in the morn
    that waked the priest all shaven and shorn
      that married the man all tattered and torn
        that kissed the maiden all forlorn
          that milked the cow with the crumpled horn
            that tossed the dog
              that worried the cat
                that killed the rat
                  that ate the malt
                    that lay in the house
                      that Jack built.

This is the house that Jack built.

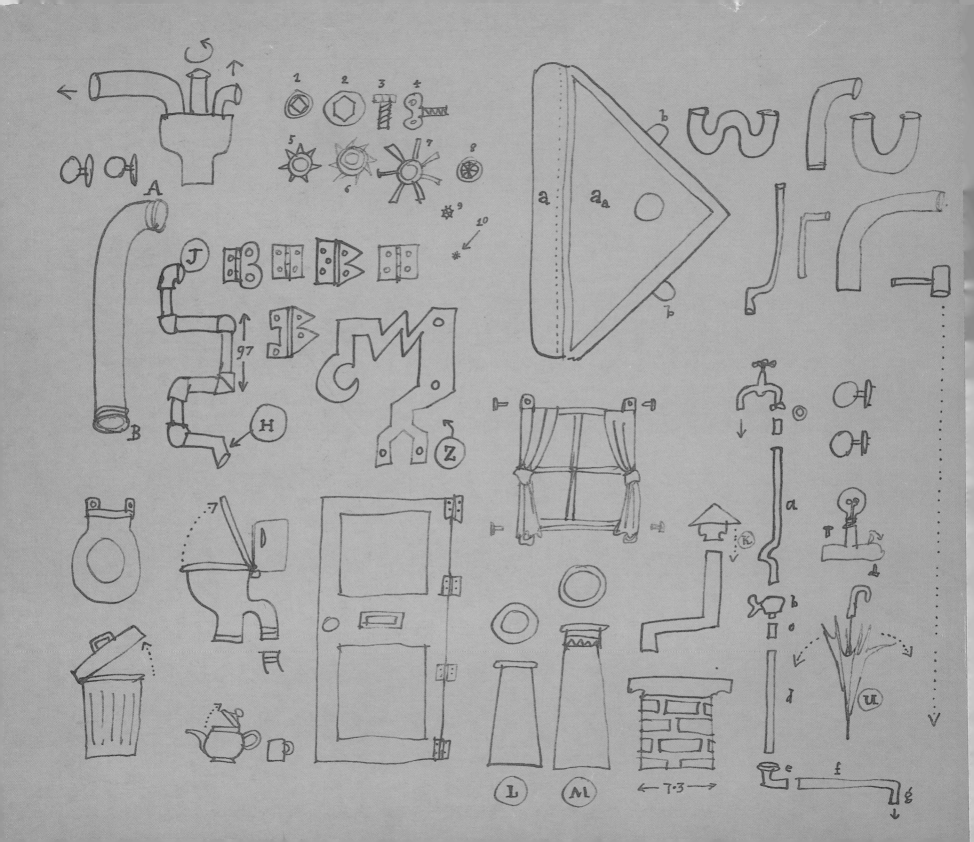